SCIENCE AnyTime™

Workbook

Teacher's Edition
UNITS E-H

HARCOURT BRACE & COMPANY

Orlando Atlanta Austin Boston San Francisco Chicago Dallas New York
Toronto London

To the Teacher

This *Science AnyTime*™ *Workbook* is designed to assist you in helping students develop their understanding of science concepts. The various exercises provide review and reinforcement of concepts covered in the units and may be especially useful to students as they prepare for tests or other assessments. You may also wish to use these exercises in an ongoing evaluation of student progress.

The workbook activities are organized by unit, section, and lesson. There is *one page for every lesson,* and a variety of formats is used to interest and engage the children. The science concept is referenced for the teacher at the bottom of each page. The unit ends with a Language of Science activity, which reinforces the vocabulary covered in the unit.

Copyright © by Harcourt Brace & Company

All rights reserved. No part of this publication may be reproduced or transmitted in any form or by any means, electronic or mechanical, including photocopy, recording, or any information storage and retrieval system, without permission in writing from the publisher.

Requests for permission to make copies of any part of the work should be mailed to: Permissions Department, Harcourt Brace & Company, 6277 Sea Harbor Drive, Orlando, Florida 32887-6777.

HARCOURT BRACE and Quill Design is a registered trademark of Harcourt Brace & Company.

Printed in the United States of America

ISBN 0-15-306796-9

1 2 3 4 5 6 7 8 9 10   085   98 97 96 95

Workbook

CONTENTS

UNIT E   Dinosaur Museum
   Section 1 • Landforms **E1**
   Section 2 • Fossils **E5**
   Section 3 • Dinosaurs **E10**
   Section 4 • Endangered Species **E18**

UNIT F   Science in the Kitchen
   Section 1 • Classifying and Measuring Matter **F1**
   Section 2 • Water Changes State **F9**
   Section 3 • Mixtures **F13**
   Section 4 • Heat **F18**

UNIT G   The Great Outdoors
   Section 1 • Habitat **G1**
   Section 2 • Desert and Rain Forest **G4**
   Section 3 • Forest **G10**
   Section 4 • Ocean **G14**
   Section 5 • Pollution Solutions **G19**

UNIT H   Show Time!
   Section 1 • Art Gallery **H1**
   Section 2 • Making Shadows **H6**
   Section 3 • Making Music 1 .... **H13**
   Section 4 • Making Music 2 .... **H18**

UNIT E: Dinosaur Museum

Section 1 • Landforms
    Lesson 1 • Canyons and Dunes **E1**
    Lesson 2 • Water Changes Rocks **E2**
    Lesson 3 • Water Changes the Land **E3**
    Lesson 4 • Another Force **E4**

Section 2 • Fossils
    Lesson 1 • Layers and Layers **E5**
    Lesson 2 • Fascinating Fossils **E6**
    Lesson 3 • It's Evident! **E7**
    Lesson 4 • Digging In **E8**
    Lesson 5 • Inferring from Footprints **E9**

Section 3 • Dinosaurs
    Lesson 1 • Dinosaur Days **E10**
    Lesson 2 • Skin Tight **E11**
    Lesson 3 • Moving Along **E12**
    Lesson 4 • Putting Bones Together **E13**
    Lesson 5 • A Toothy Grin **E14**
    Lesson 6 • Construct-O-Saurus **E15**
    Lesson 7 • Look-Alikes **E16**
    Lesson 8 • What Happened to the Dinosaurs? **E17**

Section 4 • Endangered Species
    Lesson 1 • Extinct and Endangered **E18**
    Lesson 2 • Circle of Life **E19**
    Lesson 3 • Disappearing Act **E20**
    Lesson 4 • How Can We Save the Animals? **E21**

Language of Science **E22**

Name _____

Section 1
Lesson 1
Workbook

Canyons and Dunes

Canyons and sand dunes are landforms. They are made when forces such as wind and water change the way the land looks. A force like wind can build up a sand dune. A force like water can wear down rock until a canyon is formed.

Some of these changes happen quickly. Sand can move fast when it is blown by the wind. A flood can quickly wash away land and the buildings and roads that are built on the land. But sometimes, the changes take a long time.

Wind or water changed the way the land looks in these pictures. Color the parts that wind or water changed.

The canyon walls and the sand dune should be colored.

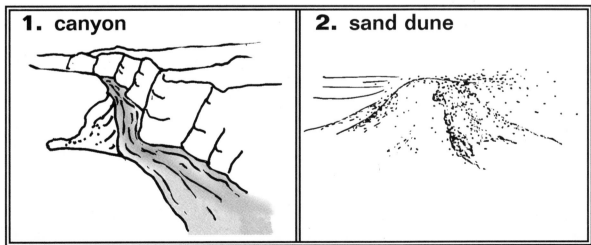

1. canyon
2. sand dune

Concept: Landforms are formed and broken down by moving water and wind. Some landforms have changed over many years.

Workbook
DINOSAUR MUSEUM
E1

Name _____

**Section 1
Lesson 2
Workbook**

Water Changes Rocks

It took many years for the Grand Canyon to be formed. The water of the Colorado River washed over the rock and began to wear it away. This formed the walls of the canyon. Wind, rain, and melting snow caused more rock to wear away.

Draw a picture of another landform that is formed by the movement of wind or water.

Drawings will vary, but should show a landform formed by the movement of wind or water.

E2 **Workbook
DINOSAUR MUSEUM**

Concept: Landforms are formed and broken down by moving water and wind. Some landforms have changed over many years.

Name _____

**Section 1
Lesson 3
Workbook**

Water Changes the Land

Wherever there is water, the land around is changed. You can see this if you look at a riverbank or an ocean shore. When the water moves over the land, it **erodes** it, or wears it away.

Rain can erode land too. When there is a lot of rain, an area can flood. The water in a flood washes away signs, cars, and even houses and roads.

Draw before and after pictures. Show how water changed something you have seen.

1. Before	2. After
Drawing should show an object before it is eroded by water.	**Drawing should show the same object after it is eroded by water.**

Concept: Landforms are formed and broken down by moving water and wind. Some landforms have changed over many years.

Workbook
DINOSAUR MUSEUM E3

Name _____

**Section 1
Lesson 4
Workbook**

Another Force

Another force that changes land is abrasion. Abrasion happens when one object rubs against another. Little pieces of the objects break off, and the objects are worn down.

When rocks rub against each other, the rocks get smaller. Tiny bits of them are worn away because of the abrasion. Sharp edges of them might get rounded. When rocks rub against each other in a stream, the rocks change in size and shape. When sand blows against rock, both the sand and the rock can be changed.

1. Draw a picture to show how this rock would change if other rocks rubbed against it for a long time.

Drawing should show a smoother, smaller rock.

2. Draw a picture to show how this rock would change if sand blew against it for a long time.

Drawing should show smoother, smaller rock.

Concept: Landforms are formed and broken down by moving water and wind. Some landforms have changed over many years.

Name _____

**Section 2
Lesson 1
Workbook**

Layers and Layers

A **fossil** is the print or hardened shape of a plant or animal. Fossils are found in rock layers. These layers of rock were once soil. Plants and animals that died became buried in the soil. Over many years, because of the pressure of all the other soil above it, the soil became rock. New layers of rock keep forming over older layers.

Draw and color some fossils in the oldest rock layer in the picture above.

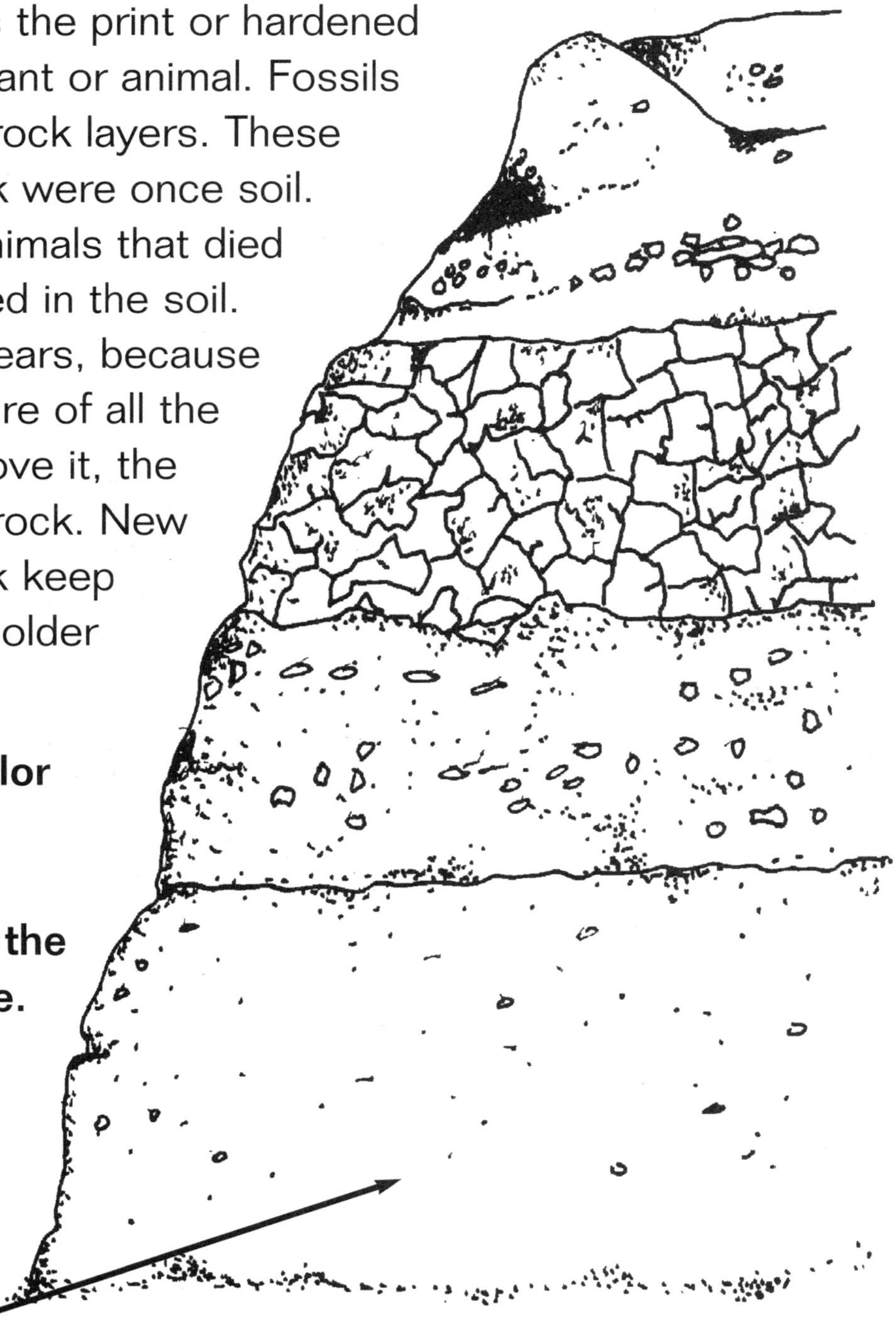

Children should draw and color some fossils in this rock layer.

Concept: Fossils are preserved remnants or impressions of organisms that lived long ago.

Workbook
DINOSAUR MUSEUM E5

Name _____

**Section 2
Lesson 2
Workbook**

Fascinating Fossils

Since different rock layers form at different times, the fossils found in the layers are different too. This helps scientists know when certain plants and animals lived. For example, fossils of starfish were found in deeper layers of rock than fossils of **dinosaurs.** Which animal lived first?

Color the oldest fossils in the picture. How do you know they are the oldest?

Fossils in this layer should be colored.

Possible response: The oldest rock layer is on the bottom, so the fossils in that layer are oldest, too.

E6 **Workbook
DINOSAUR MUSEUM**

Concept: Fossils are preserved remnants or impressions of organisms that lived long ago.

Name _____

**Section 2
Lesson 3
Workbook**

It's Evident!

One kind of fossil is the hardened print of a plant or animal. This is like a footprint you might make in the sand. Other kinds of fossils are molds and casts. A mold is made when the soil around the plant or animal becomes hard, and then the plant or animal decays. What is left is a hollow mold of its shape. If the mold is later filled in with some other material, that fossil is called a cast.

Match the word with the fossil.

1. shell

2. leaf

3. footprint

Concept: Fossils are preserved remnants or impressions of organisms that lived long ago.

Workbook
DINOSAUR MUSEUM E7

Name _____

**Section 2
Lesson 4
Workbook**

Digging In

Scientists who are interested in knowing about plants and animals that lived long ago are called **paleontologists.** One way paleontologists find out about the past is by finding and observing fossils.

Paleontologists must be very careful when they dig for fossils. They must not break the fossils they find. They often make a map to show exactly where they find the fossils.

Imagine that you are the paleontologist who found this fossil. Use the grid to draw where you found it.

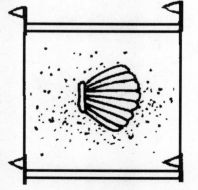 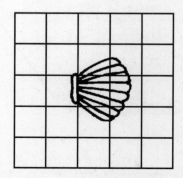

**Workbook
DINOSAUR MUSEUM**

Concept: Fossils are preserved remnants or impressions of organisms that lived long ago.

Name _____

**Section 2
Lesson 5
Workbook**

Inferring from Footprints

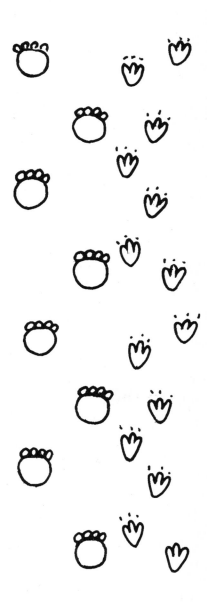

Paleontologists have learned about plants and animals that lived on Earth long before people were alive. How could they find out what happened long ago, if no person was there to see it?

Paleontologists can tell some things about the way life was millions of years ago by observing fossils. Dinosaur fossils show the size of dinosaurs, what they ate, and how they moved. When different fossils are found together, they can even tell about an event that may have taken place long ago.

Look at these fossil footprints that show two dinosaurs walking together. What do they tell you?

_____
Possible response: Two different kinds of dinosaurs were walking next to each other.
_____
_____
_____

Concept: Dinosaurs were living things that inhabited the Earth long ago.

Workbook
DINOSAUR MUSEUM E9

Name _____

**Section 3
Lesson 1
Workbook**

Dinosaur Days

Dinosaurs were reptiles that lived on the Earth millions of years ago. They looked and acted different from each other, just as animals do now. Some dinosaurs were large, and some were small. Some dinosaurs ate plants, and some ate meat. Some walked on two legs, and some walked on four legs.

Paleontologists know about dinosaurs from their observations of fossils and from what they know about living animals. While paleontologists have never seen a living dinosaur, what they know about dinosaurs is based on facts.

Draw a picture of two dinosaurs. Show how dinosaurs can be different from each other.

> **Drawings will vary, but should show two different kinds of dinosaurs.**

**Workbook
DINOSAUR MUSEUM**

Concept: Dinosaurs were living things that inhabited the Earth long ago.

Name _____

**Section 3
Lesson 2
Workbook**

Skin Tight

You have probably seen colored pictures of dinosaurs. Most artists paint them brown, gray, or green. But no one really knows what color dinosaurs were. We do know the texture of their skin, because fossils of dinosaur skin have been found.

Dinosaur skin helped dinosaurs live in their environments. Some dinosaur skin may have been brightly colored to look scary to an enemy or to attract other dinosaurs.

Color these dinosaurs to show the way you think they may have looked.

Accept all colors.

Concept: Dinosaurs were living things that inhabited the Earth long ago.

**Workbook
DINOSAUR MUSEUM E11**

Name _____

**Section 3
Lesson 3
Workbook**

Moving Along

Dinosaurs did not move the same way that lizards do. The legs of dinosaurs were straight beneath them. The legs of lizards bend outward so that their bellies drag on the ground. Look at the difference in this picture.

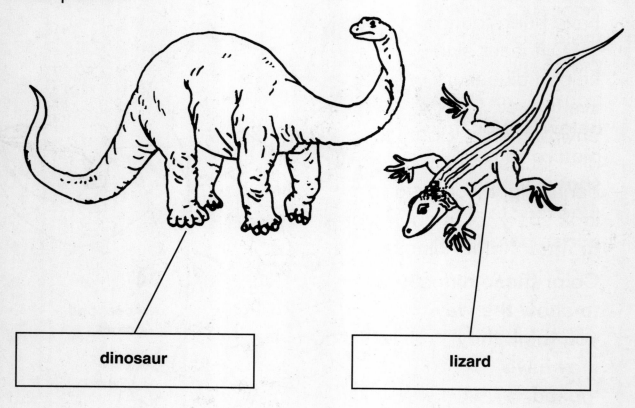

| dinosaur | lizard |

Which animal is the dinosaur?
Which animal is the lizard?

Write dinosaur or lizard in the boxes.

**Workbook
DINOSAUR MUSEUM**

Concept: Dinosaurs were living things that inhabited the Earth long ago.

Name _____

Putting Bones Together

Section 3
Lesson 4
Workbook

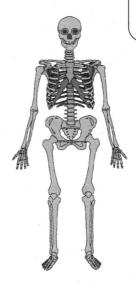

You can probably tell that the bones in this picture are the bones of a person. You can tell because the bones fit the shape of a person.

When paleontologists find dinosaur bones, they put them together. That tells them the shape of the dinosaur.

The stegosaurus should be colored.

Look at the bones below. Color the picture that shows which dinosaur these bones belong to.

Concept: Dinosaurs were living things that inhabited the Earth long ago.

Workbook
DINOSAUR MUSEUM E13

Name _____

**Section 3
Lesson 5
Workbook**

A Toothy Grin

Paleontologists are able to tell what kind of food dinosaurs ate by looking at fossils of teeth. Dinosaurs that were meat eaters needed sharp teeth to pull the meat off animals they had attacked. Dinosaurs that were plant eaters needed flat, wide teeth to grind up plants.

Draw teeth on the dinosaurs below.

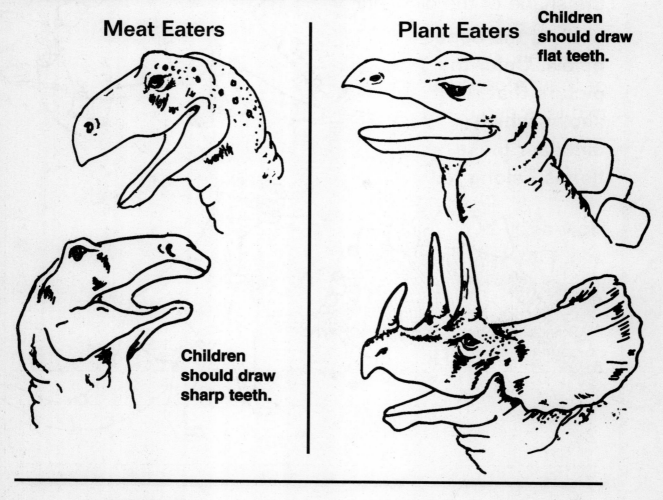

Meat Eaters — Children should draw sharp teeth.

Plant Eaters — Children should draw flat teeth.

Concept: Dinosaurs were living things that inhabited the Earth long ago.

Name _____

**Section 3
Lesson 6
Workbook**

Construct-O-Saurus

All animals have features that help them live in their habitats. These are called adaptations.

Plant-eating dinosaurs had heavy skin, and many had spikes, horns, or armor. This protected the dinosaurs. Most plant-eating dinosaurs walked on four heavy legs. This would help balance them as they ate.

Meat-eating dinosaurs had lightweight skin to keep them cool during hunting. Many of them moved on two legs and could run quickly when they were chasing prey.

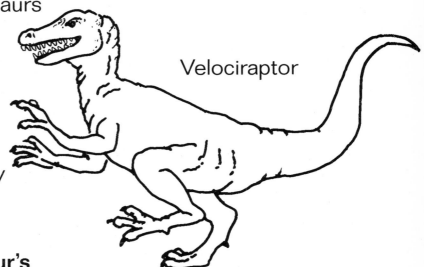
Velociraptor

How did this dinosaur's adaptations help it live in its habitat?

Possible response: This dinosaur had sharp teeth for eating meat.

_____

_____

_____

Concept: Dinosaurs were living things that inhabited the Earth long ago.

Workbook
DINOSAUR MUSEUM **E15**

Name _____

**Section 3
Lesson 7
Workbook**

Look-Alikes

Triceratops   rhinoceros

This Triceratops and this rhinoceros look alike in some ways, but they are not related to each other. Animals develop adaptations that help them stay alive in their habitats. Horns can help animals protect themselves. Some modern animals have horns. Some dinosaurs did, too.

1. Think of another animal that you think looks like a dinosaur.

2. Draw a picture of that animal and the dinosaur it looks like.

Drawings should show a modern animal and the dinosaur it resembles.

**Workbook
DINOSAUR MUSEUM**

Concept: Dinosaurs were living things that inhabited the Earth long ago.

Name _____

**Section 3
Lesson 8
Workbook**

What Happened to the Dinosaurs?

No dinosaurs live on Earth today. They are **extinct.** That means that they don't exist now and that there will never be any more of them.

Some scientists think that the dinosaurs died out from diseases. Others think that the Earth was hit by an asteroid from space or that volcanoes erupted. If either of those two things happened, so much dust might have been sent into the air that the light of the sun would have been blocked. That would have changed the weather. If dinosaurs or the things they ate couldn't live in the new weather, the dinosaurs would die out.

No one knows for sure why the dinosaurs became extinct. What do you think?

Possible response: I think the dinosaurs died out because the weather changed.

_____

_____

_____

Concept: Dinosaurs were living things that inhabited the Earth long ago.

**Workbook
DINOSAUR MUSEUM**

Name _____

Extinct and Endangered

**Section 4
Lesson 1
Workbook**

Dinosaurs are not the only animals that are extinct. In fact, many kinds of animals and plants become extinct every day.

Other animals and plants are **endangered.** This means that they are in danger of becoming extinct. They need special help and protection to survive.

There are many reasons why living things become extinct or endangered. Some animals are killed for their <u>meat</u>, <u>skin</u>, <u>feathers</u>, or <u>fur</u>. Some are killed for body <u>parts</u>, such as ivory tusks. Some are killed because people are <u>afraid</u> of them. Others are killed by <u>accident</u>.

Under the pictures, write why you think the animal is endangered. Use the underlined words.

Possible responses: skin, parts, or afraid

1. _____

Possible responses: fur, afraid, or accident

2. _____

Possible responses: meat, skin, afraid, or accident

3. _____

E18 Workbook
DINOSAUR MUSEUM

Concept: Since the time of the dinosaurs, other living things have become extinct. Some living things are endangered.

Name _____

**Section 4
Lesson 2
Workbook**

Circle of Life

Living things can also become endangered or extinct because their habitat is changed or destroyed. A **habitat** is a place where an animal or plant lives. A habitat has all the things in it that the living thing needs: food, water, air, and shelter. Habitats are changed when trees are cut down, grassland is plowed under, and swamps are drained. Air pollution, acid rain, and the use of poisons also change or destroy habitats.

1. Color this forest habitat. **The habitat should be colored.**
2. On another piece of paper, tell why these animals and plants live in this habitat.

Possible response: This habitat meets the needs of these animals and plants.

Concept: Since the time of the dinosaurs, other living things have become extinct. Some living things are endangered.

**Workbook
DINOSAUR MUSEUM**   **E19**

Name _____

**Section 4
Lesson 3
Workbook**

Disappearing Act

Giant pandas are endangered animals. There are only about 900 of them left. They live in the mountain forests of China.

Pandas are plant eaters. Their favorite food is bamboo. They eat from 22 to 44 pounds (10 to 20 kilograms) of bamboo every day. It takes them about 14 hours a day just to eat!

People used to hunt pandas for their fur. Now there are laws against this. But pandas are still in danger because they can't find enough bamboo to eat. Many of the bamboo forests have been cut down to make way for houses and fields.

People are trying to help keep pandas alive. They have made special parks where pandas can come to get extra bamboo.

Draw and color a habitat for this giant panda.

Children should draw and color a bamboo forest around the panda.

**Workbook
DINOSAUR MUSEUM**

Concept: Since the time of the dinosaurs, other living things have become extinct. Some living things are endangered.

Name _____

**Section 4
Lesson 4
Workbook**

How Can We Save the Animals?

Many people all over the world are trying to save endangered animals. To do this, people have to make a plan. The plan is different for each kind of animal. People find out why the animal is in trouble and then do what they can do to help. They know that all living things are important to every other living thing on our planet.

Why do people need to make a different plan for saving each different kind of endangered animal?

Possible response: There may be a different reason why each animal is endangered.

Concept: Since the time of the dinosaurs, other living things have become extinct. Some living things are endangered.

Workbook
DINOSAUR MUSEUM **E21**

Name _____

Unit E Workbook

Language of Science

1. Draw a picture of a **paleontologist** finding a **fossil**.

> **Drawings will vary, but should show a paleontologist finding a fossil.**

Mark an **X** in the correct space.

2. Dinosaurs are

___ endangered.

_X_ extinct.

3. How can water **erode** rock?

Possible response: Water can run over rock until the rock wears away.

_____

_____

_____

_____

E22 Workbook
DINOSAUR MUSEUM

Language of Science

UNIT F: Science in the Kitchen

Section 1 • Classifying and Measuring Matter
Lesson 1 • Observing Matter **F1**
Lesson 2 • Cutting, Bending, Twisting, Weaving **F2**
Lesson 3 • Solids, Liquids, and Gases **F3**
Lesson 4 • Measuring Solids—Length **F4**
Lesson 5 • Measuring Solids—Mass **F5**
Lesson 6 • Breaking, Slicing, Molding, Mixing **F6**
Lesson 7 • No Shape of Its Own **F7**
Lesson 8 • Invisible Gas **F8**

Section 2 • Water Changes State
Lesson 1 • Wonderful Water **F9**
Lesson 2 • Water Mass **F10**
Lesson 3 • Disappearing Water **F11**
Lesson 4 • Going, Going, Gone **F12**

Section 3 • Mixtures
Lesson 1 • A Kitchen Mix **F13**
Lesson 2 • Mix It Up **F14**
Lesson 3 • Take It Apart **F15**
Lesson 4 • Now You Don't See It, Now You Do **F16**
Lesson 5 • A Chilly Mix **F17**

Section 4 • Heat
Lesson 1 • Heating Things Up **F18**
Lesson 2 • Using a Thermometer **F19**
Lesson 3 • Melt Down **F20**
Lesson 4 • Warm as Toast **F21**
Lesson 5 • Popping Up a Storm **F22**
Lesson 6 • Let's Cook! **F23**
Lesson 7 • Cleaning Up the Kitchen **F24**

Language of Science **F25–F26**

Name _____

**Section 1
Lesson 1
Workbook**

Observing Matter

Everything in the world is made of **matter**. Matter is anything that takes up space and has mass. By observing matter, we can see its properties.

A **property** is a characteristic of an object. For example, the color yellow is a property of a banana. Sharpness is a property of scissors. Roundness is a property of an orange. All matter has properties.

Children should sort the objects in their own ways. Possible groupings: hard objects, sharp objects, liquids.

Sort the matter in this picture by shape or some other property. Then color the picture. Use a different color for each property.

Concept: Matter has properties that can be observed. Matter can change shape, but its essential properties remain the same.

Workbook
SCIENCE IN THE KITCHEN **F1**

Name _____

**Section 1
Lesson 2
Workbook**

Cutting, Bending, Twisting, Weaving

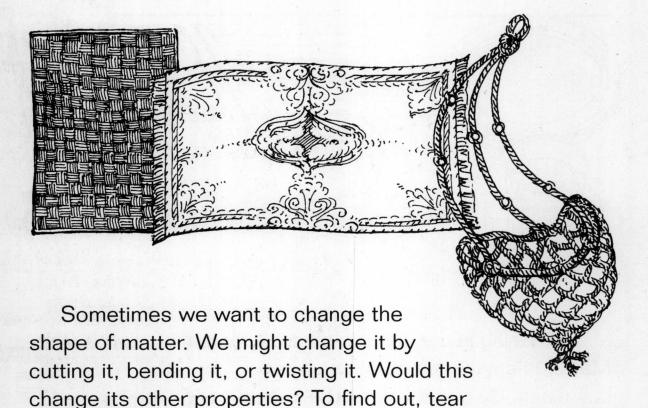

Sometimes we want to change the shape of matter. We might change it by cutting it, bending it, or twisting it. Would this change its other properties? To find out, tear a corner off of this sheet of paper. You have changed the paper's shape.

Circle yes or no to answer these questions.

1. Is the sheet of paper still paper? (yes) no
2. Has its color changed? yes (no)
3. All together, is there still the same amount of paper? (yes) no
4. Can you change the shape of matter without changing its other properties? (yes) no

**Workbook
SCIENCE IN THE KITCHEN**

Concept: Matter has properties that can be observed. Matter can change shape, but its essential properties remain the same.

Name _____

**Section 1
Lesson 3
Workbook**

Solids, Liquids, and Gases

Matter can be a **liquid** like water, a **solid** like wood, or a **gas** like air. By observing an object carefully, we can decide its properties and what kind of matter it is.

We can see a solid, and it holds its shape. We can usually see a liquid, and its shape changes easily. A gas is usually invisible, and it changes shape easily.

red: pictures, lamp, furniture, fish, bookshelf, trophies
blue: water in glass, water in aquarium

Color the solids in this picture red. Color the liquids blue. Do not color the gases.

Concept: Matter has properties that can be observed. Matter can change shape, but its essential properties remain the same.

**Workbook
SCIENCE IN THE KITCHEN**

Name _____

**Section 1
Lesson 4
Workbook**

Measuring Solids—Length

Matter can be measured. We can **measure** its **length,** or how long it is, and its **height,** or how tall it is.

Long ago, people measured objects by comparing them to something else. Many times they used parts of their bodies to measure against. For example, they might cut a piece of rope three hands long. But since the lengths of people's hands are different, they found out that this was not a very accurate way to measure. If someone else cut the rope for them, it would be a different length. Now we measure objects in units that are the same size everywhere. Then everyone knows exactly what size something is.

1. Measure the rope with your finger. Ask a friend to do the same thing. Are the two measurements the same?

   Possible response: 12; no

2. Measure some other objects with your friend.

F4 Workbook
SCIENCE IN THE KITCHEN

Concept: Matter has properties that can be observed. Matter can change shape, but its essential properties remain the same.

Name _____

**Section 1
Lesson 5
Workbook**

Measuring Solids—Mass

We can measure the **mass** of matter. Mass means how much matter is in something. One way to measure mass is to **balance** one object with another object. When the balance tips, you know which object has more mass and which object has less mass.

Circle the objects that have more mass in each picture. They may not always be the biggest objects!

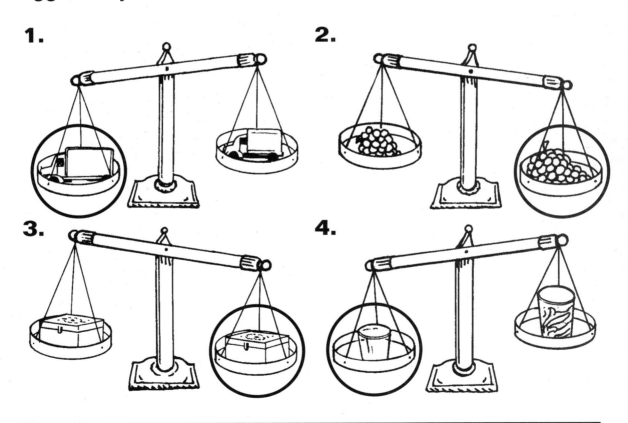

Concept: Matter has properties that can be observed. Matter can change shape, but its essential properties remain the same.

Workbook
SCIENCE IN THE KITCHEN

Name _____

**Section 1
Lesson 6
Workbook**

Breaking, Slicing, Molding, Mixing

You have probably made things from clay. You can break a lump of clay into pieces, slice it, or mold it. You can change its shape over and over. But it is still clay. Unless you throw away some clay, it will have the same mass, too.

The same thing is true about all kinds of matter. If you change the shape of the matter but keep all of the matter together when you measure it, its mass stays the same.

1. Draw in the box something you could make from the clay.

2. Is the mass of your object different from the mass of the lump of clay?   no

Drawing should show an object made from clay.

F6 Workbook
SCIENCE IN THE KITCHEN

Concept: Matter has properties that can be observed. Matter can change shape, but its essential properties remain the same.

Name _____

**Section 1
Lesson 7
Workbook**

No Shape of Its Own

Water is a liquid. A liquid changes shape easily. If you pour water from a tall glass into a bowl, the water has changed in shape. But there is still just as much water as there was before. You could prove this by measuring the water with a measuring cup. You would be measuring the **volume** of water, or the amount of space the water takes up.

This picture shows 1 cup of water. Draw the same water in 3 other shapes.

1.	2.	3.
Children should draw the water in three other shapes.		

4. What is the volume of water in each shape?
_____

1 cup

Concept: Matter has properties that can be observed. Matter can change shape, but its essential properties remain the same.

**Workbook
SCIENCE IN THE KITCHEN F7**

Name _____

**Section 1
Lesson 8
Workbook**

Invisible Gas

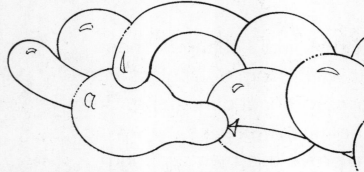

Is an empty cup <u>really</u> empty? What about an empty pocket? Or an empty lunch box? All of these things are really full. They are full of air.

Air is an invisible gas. Since we can't see it, we sometimes forget it is there. But it takes up space.

Air takes the shape of the container it is in. Think of a thin balloon and a round balloon. Since the balloons are different shapes, the air inside is different shapes.

Write a sentence about air.

　　**Possible response: Air takes the
　　shape of its container.**
_____
_____
_____
_____

Concept: Matter has properties that can be observed. Matter can change shape, but its essential properties remain the same.

Name _____

Section 2
Lesson 1
Workbook

Wonderful Water

Water is an interesting kind of matter. It can be a solid, a liquid, or a gas. If water is a solid, it is ice. We can drink liquid water or use it in other ways. Water that is a gas is called water vapor.

Water can change from one form to another. If you freeze liquid water, it becomes a solid. If you melt the solid ice, it becomes a liquid. If you leave a glass of liquid water out without covering it, it **evaporates** and becomes a gas. This gas is invisible and is part of the air.

Draw lines to match the words on the left with the words on the right.

1. ice — liquid
2. water vapor — solid
3. water — gas

Concept: Water can be a solid, liquid, or gas and can change from one form to another.

Name _____

**Section 2
Lesson 2
Workbook**

Water Mass

Since water is matter, we know that it has mass. Do you think liquid water has the same mass as frozen water? To compare the two forms of water, we could use a balance. If the balance tips, we know that one form of water has more mass than the other. If the balance stays level, both forms of water have the same mass.

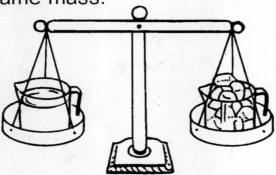

This picture shows a balance. There is one cup of liquid water on one side and one cup of water that has been frozen on the other side.

What do you observe about the mass of frozen water?

Possible response: One cup of frozen water and one cup of liquid water have the same mass.

F10 Workbook
SCIENCE IN THE KITCHEN

Concept: Water can be a solid, liquid, or gas and can change from one form to another.

Name _____

**Section 2
Lesson 3
Workbook**

Disappearing Water

Water changes in form all of the time. We see liquid water in rivers, lakes, and oceans. Some of that water evaporates because of the wind and the heat of the sun. It changes into water vapor. When the water vapor condenses, it forms a cloud. When the droplets in the cloud become too heavy, they fall as rain. The same thing happens over and over. This is called the water cycle.

Add labels to this picture of the water cycle. Use the words in the box.

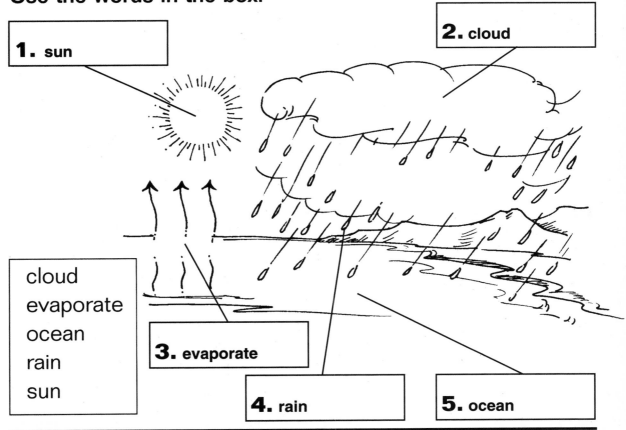

1. sun
2. cloud
3. evaporate
4. rain
5. ocean

cloud
evaporate
ocean
rain
sun

Concept: Water can be a solid, liquid, or gas and can change from one form to another.

Workbook
SCIENCE IN THE KITCHEN **F11**

Name _____

**Section 2
Lesson 4
Workbook**

Going, Going, Gone

Have you ever put wet clothes in a dryer or hung them on a clothesline? You might do those things if you want the clothes to get dry. Heat and moving air make water evaporate more quickly than cool, still air.

These pictures both show wet clothes on a clothesline. In the first picture, the sky is cloudy and the air is still. In the second picture, the day is sunny and the wind is blowing.

Color the picture that shows clothes that will dry faster.

This picture should be colored.

F12 **Workbook
SCIENCE IN THE KITCHEN**

Concept: Water can be a solid, liquid, or gas and can change from one form to another.

Name _____

**Section 3
Lesson 1
Workbook**

A Kitchen Mix

Have you ever made a mixture? You may have mixed a vegetable salad or fruit salad. You may have mixed up some clothes in a drawer. A **mixture** is made when you combine several things.

The things that go into a mixture can be separated. You can take the different vegetables or fruits out of a salad. They will be the same as they were before. You can sort out the clothes in the drawer. They will not have changed.

Separate the things in these mixtures. Draw them in different groups.

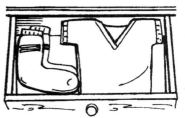

1. Drawing should show groups of carrots, tomatoes, and lettuce.

2. Drawing should show groups of apple slices, orange slices, and grapes.

3. Drawing should show groups of socks and T-shirts.

Concept: Matter can be mixed. The individual properties that make up the mixture do not change.

Name _____

**Section 3
Lesson 2
Workbook**

Mix It Up

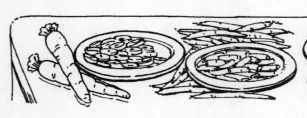

This boy is helping to make vegetable soup for his family. He will mix different vegetables together and then mix in salt and flavorings.

1. What could someone in his family do if they did not want to eat the carrots in the soup?

> **Possible response: They could take the carrots out.**

Draw pictures of two other mixtures that people could eat or drink.

2. **Children should draw two mixtures.**

3.

F14 Workbook
SCIENCE IN THE KITCHEN

Concept: Matter can be mixed. The individual properties that make up the mixture do not change.

Name _____

**Section 3
Lesson 3
Workbook**

Take It Apart

There are different ways that things in a mixture can be separated.

This is a mixture of rocks. You could separate the different kinds of rocks with your hands.	
This is a mixture of gravel and sand. You could separate the gravel from the sand with a screen.	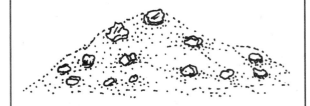
This is a mixture of metal and plastic figures. You could separate the metal figures from the plastic ones with a magnet.	
This is a mixture of coffee grounds and water. You could separate the coffee grounds from the water with a filter.	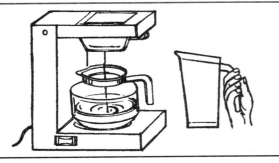

1. On another sheet of paper, draw a mixture.

2. Show how you would separate it.

Children should draw a mixture and show how it can be separated.

Concept: Matter can be mixed. The individual properties that make up the mixture do not change.

**Workbook
SCIENCE IN THE KITCHEN F15**

Name _____

**Section 3
Lesson 4
Workbook**

Now You Don't See It, Now You Do

If you have ever made salt water, you've made another kind of mixture. The salt is mixed into the water until it has **dissolved**. When something dissolves, it seems to disappear. It is still there, but the pieces of it are so small that they can't be seen.

Just like in mixtures of other things, the salt can be separated from the water. To do it, you would need to let the water evaporate. The salt would be left behind.

This is a picture of lemonade. Lemonade is a mixture of lemon juice, water, and sugar.

Write a sentence that tells a way to get the sugar out of lemonade.

Possible response: A way to get sugar out of lemonade is to let the water evaporate.

F16   **Workbook
SCIENCE IN THE KITCHEN**

Concept: Matter can be mixed. The individual properties that make up the mixture do not change.

Name _____

**Section 3
Lesson 5
Workbook**

A Chilly Mix

It took about 500 years to get ice cream from China to America! An explorer named Marco Polo learned about ice cream when he was visiting China. He brought a recipe to Italy. From there it traveled to France and the rest of Europe.

An American named Thomas Jefferson was in France and learned how to make ice cream. When he came back to America, he taught people here how to make it.

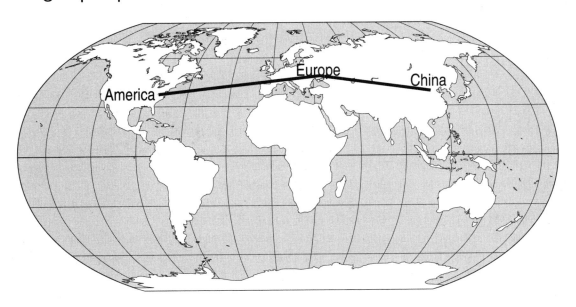

With a red crayon, draw a line that shows how ice cream traveled to America.

Concept: Matter can be mixed. The individual properties that make up the mixture do not change.

Workbook
SCIENCE IN THE KITCHEN **F17**

Heating Things Up

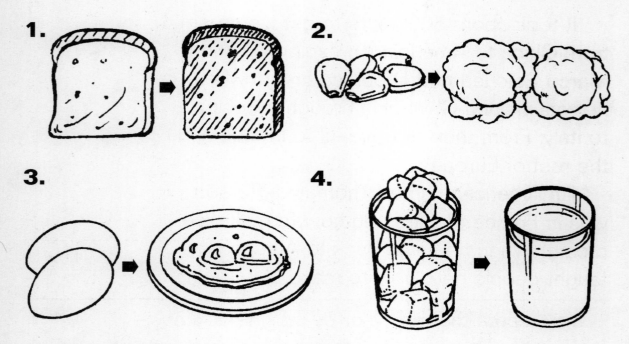

People often use heat when they cook. **Heat** is a form of energy that can change matter. If you heat ice, it will change the solid ice to liquid water. If you heat an egg, the egg will become a solid. If you heat bread by toasting it, the bread will turn brown and get harder.

Heat makes some changes that can be reversed, or changed back. Some of the changes can not be reversed.

Color the pictures that show the matter that has been changed by heat.

The toast, popcorn, cooked eggs, and liquid water should be colored.

Name _____

**Section 4
Lesson 2
Workbook**

Using a Thermometer

When you want to know how hot or cold matter is, you want to find out its **temperature.** You can use a **thermometer** to measure temperature. The liquid in a thermometer goes up when it is hot and down when it is cold. Which way would the liquid move if you put a thermometer into a freezer?

Possible response: The liquid would go down.

You read a thermometer by looking at the markings on it. The markings tell the number of degrees the thermometer is measuring.

Write the number of degrees these thermometers show.

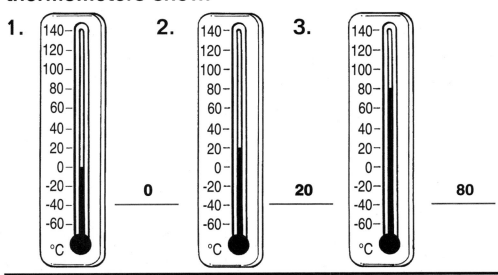

1. 0   2. 20   3. 80

Concept: Heat affects matter.

Name _____

Section 4
Lesson 3
Workbook

Melt Down

Heat moves from something warm to something cold. Hot soup will cool off in a kitchen. The heat from the soup will move into the air until the soup and the air are the same temperature. Ice cream will melt in a kitchen. Since the temperature of the kitchen air is warmer than the ice cream, the heat from the kitchen air melts the ice cream. The heat keeps moving until the melted ice cream and the air are both the same temperature.

On a cold day, you might be asked to keep the windows and outside doors of your home shut. If it is a hot day, you might be asked to do the same thing. Why? **Use another sheet of paper to write your answer.**

Possible response: On either day, heat would move through the open doors and windows to the cooler place.

F20 Workbook
SCIENCE IN THE KITCHEN

Concept: Heat affects matter.

Name _____

**Section 4
Lesson 4
Workbook**

Warm as Toast

When heat changes something and it can't be changed back, there is chemical change. If you add heat to popcorn kernels, they will turn into popcorn. Could you change the popcorn back to kernels again? If you add heat to bread, the bread turns into toast. Could you change the toast back into bread?

1. Draw what will happen if you add heat to these foods.

2. Circle the change that can be reversed.

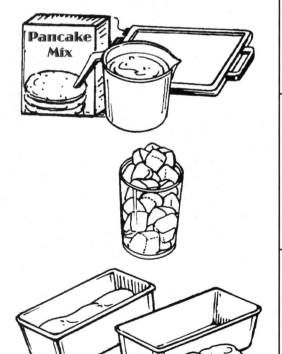

1.
Drawing should show pancakes.

2.
Drawing should show liquid water.
Drawing should be circled.

3.
Drawing should show bread.

Concept: Heat affects matter.

Workbook
SCIENCE IN THE KITCHEN **F21**

Name _____

**Section 4
Lesson 5
Workbook**

Popping Up a Storm

Do you like to eat corn? Corn is a food that can be ground, mashed, cooked, or mixed with other things to make many different foods.

Make a list of four foods that are made with corn.

Possible responses:

1. popcorn, corn muffins, corn oil, tortillas, corn flakes, corn grits

2. _____

3. _____

4. _____

Corn is used in more ways than just for food. Parts of the corn plant are used to make toothpaste, glue, crayons, soap, and many other things. We would probably find it hard to get along without corn!

**Workbook
SCIENCE IN THE KITCHEN**

Concept: Heat affects matter.

Name _____

**Section 4
Lesson 6
Workbook**

Let's Cook!

People prepare food in many different ways. Sometimes they use the <u>heat</u> from a stove or fire to cook it. Sometimes they use a <u>cold</u> refrigerator or freezer to cool it. Some foods are eaten <u>whole</u>, and some foods are <u>chopped</u>, <u>ground</u>, or <u>mashed</u>. Sometimes food is <u>mixed</u> with other food before we eat it.

1. Draw a meal that you would like to make and eat.

2. Next to the plates and glass, write how you changed matter to make your meal. Use the underlined words.

Children should write some of the underlined words next to the plates and glass.

Drawing should show food on the plates and a drink in the glass.

Concept: Heat affects matter.

Workbook
SCIENCE IN THE KITCHEN **F23**

Name _____

**Section 4
Lesson 7
Workbook**

Cleaning Up the Kitchen

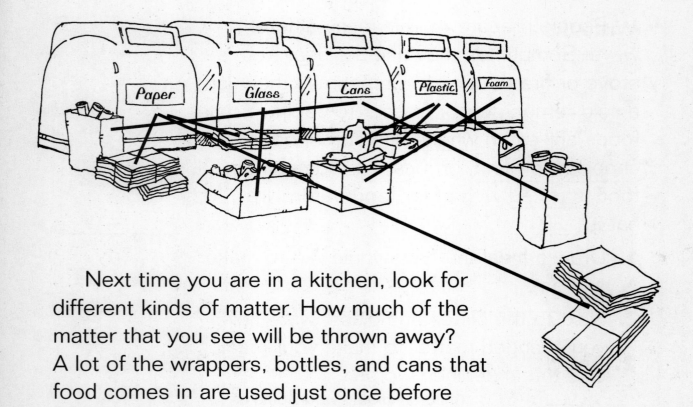

Next time you are in a kitchen, look for different kinds of matter. How much of the matter that you see will be thrown away? A lot of the wrappers, bottles, and cans that food comes in are used just once before they are put in the trash. This wastes the Earth's resources. Now people are trying to recycle things like paper, glass, cans, plastic, and foam. **Recycling** means using the matter that things are made of to make new things. Recycling will help protect our environment.

In the picture above, draw a line from each object to the bin it should be put into for recycling.

**Workbook
SCIENCE IN THE KITCHEN**

Concept: Heat affects matter.

Name _____

Unit F Workbook

Language of Science

Write the correct word for each space.

1. __**Matter**__ can be a __**solid**__,
   Matter Measure   height solid

   a __**liquid**__, or a __**gas**__.
   mass liquid   gas property

2. The __**height**__, __**length**__,
   property height   length matter

   __**mass**__, and __**volume**__
   mass solid   matter volume

   of matter can be __**measured**__.
   gas measured

Language of Science (Page 1 of 2)   Workbook **SCIENCE IN THE KITCHEN** **F25**

Name _____

Unit F Workbook

3. Draw a mixture.

Drawing should show a mixture.

4. Draw something that dissolves.

Drawing should show a substance that dissolves.

Draw a line to the word that answers the question.

5. What is a kind of energy? ——————— heat
6. What can show us the temperature? — property
7. What is a characteristic of something? — thermometer

Draw a line to match each package with its recycling bin.

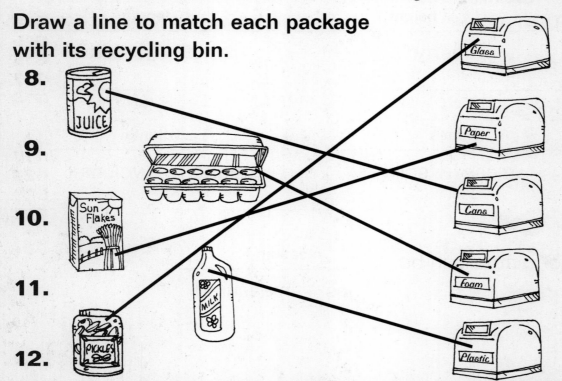

8.
9.
10.
11.
12.

F26 Workbook
SCIENCE IN THE KITCHEN   Language of Science (Page 2 of 2)

UNIT G: The Great Outdoors

Section 1 • Habitat
    Lesson 1 • Looking at My World **G1**
    Lesson 2 • Up Close **G2**
    Lesson 3 • Making a Home **G3**

Section 2 • Desert and Rain Forest
    Lesson 1 • Wet or Dry? **G4**
    Lesson 2 • Desert Hide-and-Seek **G5**
    Lesson 3 • This Place Is Dry! **G6**
    Lesson 4 • Rain Forest Hide-and-Seek **G7**
    Lesson 5 • This Place Is Wet! **G8**
    Lesson 6 • Here or There **G9**

Section 3 • Forest
    Lesson 1 • Living Through Changes **G10**
    Lesson 2 • Forest Friends **G11**
    Lesson 3 • Terrific Trees **G12**
    Lesson 4 • Tree-mendously Useful **G13**

Section 4 • Ocean
    Lesson 1 • All Over the Place **G14**
    Lesson 2 • A Salty World **G15**
    Lesson 3 • Underwater World **G16**
    Lesson 4 • Seafood Search **G17**
    Lesson 5 • 3-D Underwater Mural **G18**

Section 5 • Pollution Solutions
    Lesson 1 • Trouble in the Great Outdoors **G19**
    Lesson 2 • Packing It Up **G20**
    Lesson 3 • Second Life **G21**
    Lesson 4 • Over and Over—Making Paper **G22**

Language of Science **G23–G24**

Name _____

**Section 1
Lesson 1
Workbook**

Looking at My World

A **habitat** is the natural home of a plant or animal. It is the place where a plant or an animal lives and grows on its own. Oceans, deserts, forests, and rain forests are habitats. Many plants and animals live in each of these habitats.

Draw a line from each animal to its habitat.

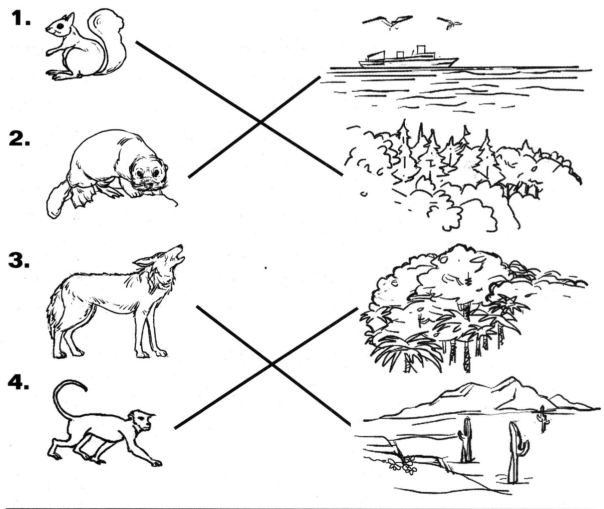

Concept: A habitat is a place where plants and animals live and that meets their needs.

**Workbook
THE GREAT OUTDOORS   G1**

Name _____

**Section 1
Lesson 2
Workbook**

Up Close

The grass beneath your feet is also a habitat. Some tiny animals make their homes there. Even the soil beneath the grass is a habitat. What animals live in the grass? What animals live beneath the grass?

A pond is also a habitat. It is the natural home of many plants and animals. What animals might you find at a pond?

Draw a line from each animal to its habitat.

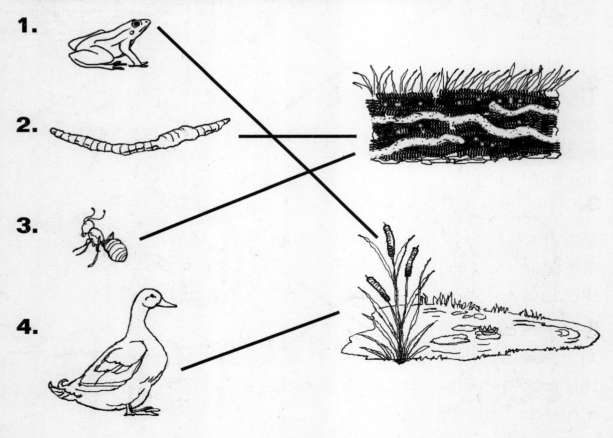

G2 Workbook
THE GREAT OUTDOORS

Concept: A habitat is a place where plants and animals live and that meets their needs.

Name _____

**Section 1
Lesson 3
Workbook**

Making a Home

All animals need three things to live. They need food to eat. They need water to drink or to take in with their food. And they need air to breathe. Each animal lives in a habitat that has the things it needs to live.

The squirrel lives in a forest. It eats nuts and seeds for food. It drinks water from the stream. It breathes the air that blows through the trees. The squirrel can get all the things it needs in the forest.

The squirrel in the picture must have three things to live. Color the three things it needs from its habitat.

The acorns, water, and air should be colored.

Concept: A habitat is a place where plants and animals live and that meets their needs.

**Workbook
THE GREAT OUTDOORS**

Name _____

**Section 2
Lesson 1
Workbook**

Wet or Dry?

This is a **desert.** The hot sun heats up the dry sand and rocks. There is little or no water around because it doesn't rain much in a desert. Few plants can grow there.

desert

This is a **rain forest.** It rains almost every day in a rain forest. Water drips from the leaves of plants. It flows in rivers and streams. Many plants grow in this habitat.

rain forest

Circle the word that tells about each habitat.

1. desert   (dry)   wet
2. rain forest   dry   (wet)

**Workbook
THE GREAT OUTDOORS**

Concept: A habitat is a place where plants and animals live and that meets their needs.

Name _____

Section 2
Lesson 2
Workbook

Desert Hide-and-Seek

Some animals live in deserts. They often hide from the hot sun. Small ones crawl under rocks. Some go into holes in the ground. Coyotes and other large animals hide in caves. They rest during the day. Then at night they hunt for food. They get some of the water they need from the plants and animals they eat.

What is wrong with the picture below? Mark an X on each animal that does not live in a desert.

Concept: A desert and a rain forest have differing climates, features, and living things.

Workbook
THE GREAT OUTDOORS
G5

Name _____

**Section 2
Lesson 3
Workbook**

This Place Is Dry!

There is little water for plants in the desert. Desert plants store water in their stems. Then they can get the water they need, even when it doesn't rain for a long time.

The **cactus** plant grows in the desert. It has a thick skin that protects it from the hot weather. It also has sharp needles that keep some animals from eating it.

The three cacti should be colored green and brown.

Find the cacti in the picture. Then color their stems green and their needles brown.

G6 Workbook
THE GREAT OUTDOORS

Concept: A desert and a rain forest have differing climates, features, and living things.

Name _____

**Section 2
Lesson 4
Workbook**

Rain Forest Hide-and-Seek

Rain forests are disappearing. Why? Because people are cutting down the trees. Other people are trying to stop the cutting. They want to save the trees and brightly colored flowers. They want to save the monkeys, parrots, and snakes. They care about all the other plants and animals that live there, too.

Color each animal that lives in the rain forest.

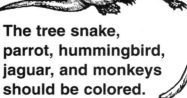

The tree snake, parrot, hummingbird, jaguar, and monkeys should be colored.

Concept: A desert and a rain forest have differing climates, features, and living things.

Workbook
THE GREAT OUTDOORS
G7

This Place Is Wet!

**Section 2
Lesson 5
Workbook**

Name _____

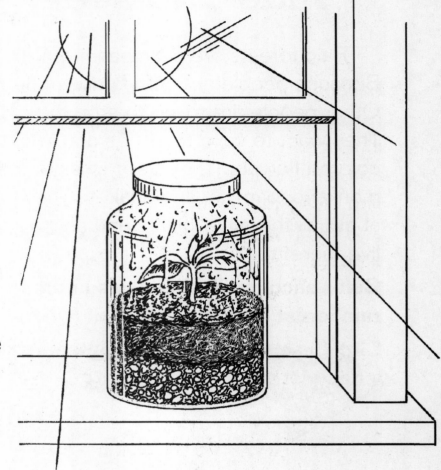

This is Jane's plant. One day after she planted it, she saw drops of water on the inside of the jar. "Where did this water come from?" Jane asked.

"This is what happens," said Jane's teacher. "The plants take up water from the soil. The plants give off some water into the air. Water also moves from the soil into the air. The lid on the jar keeps the air inside. Then the sun warms the air in the jar. And the water comes back out of the air in little drops. So the plants can use the same water over and over again."

The plant and the soil should be circled.

Where did the water on the side of the jar come from? Circle two things in the picture to show your answer.

G8 Workbook
THE GREAT OUTDOORS

Concept: A desert and a rain forest have differing climates, features, and living things.

Name _____

**Section 2
Lesson 6
Workbook**

Here or There

A monkey likes to eat soft fruits like bananas. A desert does not have these fruits. That is why the monkey does not live in a desert. It lives in a rain forest. There it can find the foods it likes to eat. And it can find plenty of water to drink, too.

Rain forest animals usually cannot live in a desert. Desert animals usually cannot live in a rain forest. Each plant and animal lives in a habitat that meets its own needs.

Circle each plant and animal that grows in a hot, wet habitat.

Pictures 1, 3, 4, and 5 should be circled.

1.

2.

3.

4.

5.

6.

Concept: A desert and a rain forest have differing climates, features, and living things.

Workbook
THE GREAT OUTDOORS

Name _____

Section 3
Lesson 1
Workbook

Living Through Changes

Trees change as seasons change. Read on to find out how a maple tree changes. Then color the pictures to show how a maple tree looks in each season.

Children should color the trees according to the descriptions.

spring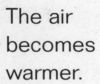

The air becomes warmer. Buds open. Little leaves begin to grow on the tree.

summer

The air grows warmer still. Big, green leaves now cover all the branches of the tree.

fall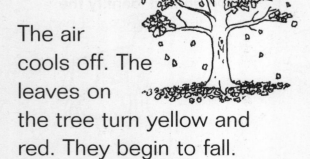

The air cools off. The leaves on the tree turn yellow and red. They begin to fall.

winter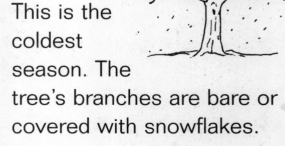

This is the coldest season. The tree's branches are bare or covered with snowflakes.

G10 Workbook
THE GREAT OUTDOORS

Concept: The boreal/deciduous forest is characterized by its climate and living things.

Name _____

Section 3
Lesson 2
Workbook

Forest Friends

broad leaves

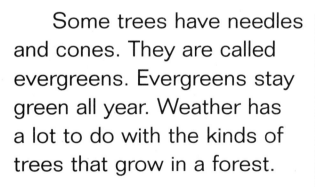

A **forest** is a place where many trees grow. Some trees have broad leaves. Trees with broad leaves are called deciduous trees.

needles and cone

Some trees have needles and cones. They are called evergreens. Evergreens stay green all year. Weather has a lot to do with the kinds of trees that grow in a forest.

1. In the box, draw a forest tree with an animal in it.
2. Circle the kind of tree that you drew.

   deciduous   evergreen

Drawing should show a deciduous or an evergreen tree with an animal in it. Child should identify the tree by circling deciduous or evergreen.

Concept: The boreal/deciduous forest is characterized by its climate and living things.

Workbook
THE GREAT OUTDOORS
G11

Name _____

Section 3
Lesson 3
Workbook

Terrific Trees

A tree takes in air through its leaves and gives off water and oxygen through its leaves. The leaves make food for the tree, too. The leaves of an evergreen are its needles.

Draw a line to match the tree to its leaves.

Ferns, wild flowers, and mushrooms grow in forests, too. Each plant grows where it can get the sunlight it needs.

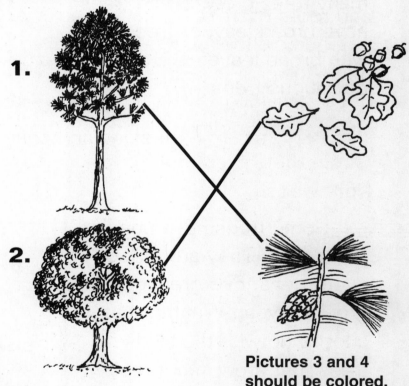

Pictures 3 and 4 should be colored.

Which things live in a forest? Color them.

3.

4.

5.
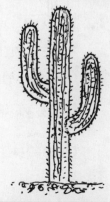

G12 **Workbook**
THE GREAT OUTDOORS

Concept: The boreal/deciduous forest is characterized by its climate and living things.

Name _____

Section 3 Lesson 4 Workbook

Tree-mendously Useful

Long ago, Native Americans made bowls, cups, and arrows from wood. They used wood to make their homes and canoes. Native Americans also used leaves, bark, and roots of trees to make medicines.

Today, people still use trees to make many things. Trees provide lumber for desks, chairs, pencils, and parts of buildings. Trees are also ground up to make paper.

Circle the things that people make from wood.

The house frame, telephone pole, pencil, paper, blocks, and picnic table should be circled.

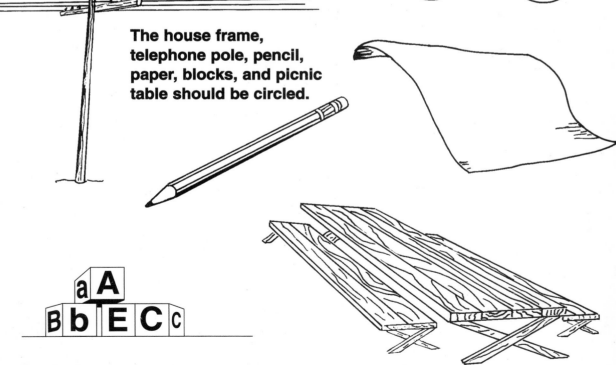

Concept: The boreal/deciduous forest is characterized by its climate and living things.

Workbook
THE GREAT OUTDOORS
G13

Name _____

**Section 4
Lesson 1
Workbook**

All Over the Place

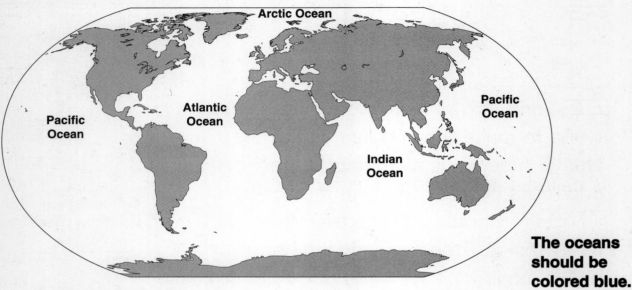

The oceans should be colored blue.

Most of the Earth's surface is covered by oceans. **Oceans** are huge bodies of water. Animals and plants live in the oceans.

The largest oceans are the Atlantic, the Pacific, and the Indian. Water flows from one ocean into another. There is one thousand times more space for animals and plants to live in oceans than on land or in the air.

1. Color the Earth's oceans blue in the picture above.
2. Now color the circle. Color three of the parts blue. Color the other part brown. Your circle should show that about $\frac{3}{4}$ of the Earth is covered by oceans.

Three parts of the circle should be colored blue. One part should be colored brown.

G14 Workbook
THE GREAT OUTDOORS

Concept: Oceans are characterized by salt water and marine plants and animals.

Name _____

**Section 4
Lesson 2
Workbook**

A Salty World

The water in oceans is salty. This water is called **salt water.** But water in lakes, ponds, rivers, and streams is not salty. It is called **fresh water.**

A change of habitat can harm a plant or animal. Sea plants and animals need the salt water of an ocean to live. They could not live in the fresh water of a lake or pond. And the plants and animals in a lake or pond could not live in the salty ocean.

Draw a picture of a water animal that could <u>not</u> live in fresh water.

Drawing should show an ocean animal, such as a shark, whale, or octopus.

Concept: Oceans are characterized by salt water and marine plants and animals.

**Workbook
THE GREAT OUTDOORS G15**

Name _____

**Section 4
Lesson 3
Workbook**

Underwater World

Seaweed and kelp grow in sunny ocean waters. These plants make good hiding places for small animals. But plants do not grow in the deepest parts of the oceans. Plants need sunlight. It is too dark at the bottom of the oceans for plants to grow. Only a few strange-looking animals live there. The lantern fish is one of them. It makes its own light.

The water near the surface should be light-colored. The water should be colored darker as the water deepens.

Color the picture to show that the water near the surface of the ocean is light-colored. Show that the color of the water becomes darker as you go deeper.

G16 Workbook
THE GREAT OUTDOORS

Concept: Oceans are characterized by salt water and marine plants and animals.

Name _____

Section 4
Lesson 4
Workbook

Seafood Search

The oceans are important to people. Many kinds of food come from ocean waters. Some seafoods that people eat are fish, oysters, crabs, and shrimps. Even seaweed and kelp are foods that some people eat.

Circle each seafood that people eat.

1. seaweed

2. crab

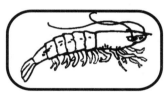
3. shrimp

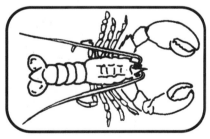
4. lobster

5. coral

6. tuna

Write the letter of the correct answer in the space on the left.

__A__ 7. Seaweed and kelp do not grow where ocean water is too _____.
    A. dark   **B.** sunny   **C.** shallow

Concept: Oceans are characterized by salt water and marine plants and animals.

Workbook
THE GREAT OUTDOORS

Name _____

**Section 4
Lesson 5
Workbook**

3-D Underwater Mural

Sharks and sting rays swim through ocean waters. Lobsters and crabs creep about on ocean floors.

Some sea animals eat plants for food. Some eat other animals. Sea animals get their food in different ways. Sharks use their sharp teeth. Jellyfish and octopuses use their tentacles. Crabs and lobsters use their claws. Sting rays and sea anemones use their stingers.

1. Find the sea animals that are numbered 1 to 7.

2. Mark an X on the part that each animal uses to catch its food.

G18 Workbook
THE GREAT OUTDOORS

Concept: Oceans are characterized by salt water and marine plants and animals.

Name _____

**Section 5
Lesson 1
Workbook**

Trouble in the Great Outdoors

Pollution is something that spoils a habitat. A soft-drink can tossed out a car window is pollution. A tire in a stream is pollution. Empty bags, plastic cups, and bottle caps are pollution, too. Even fishing line is pollution. So is a used paper diaper.

People cause pollution problems. People can solve pollution problems, too. They help by cleaning up their own trash wherever they are. They also help by cleaning up other people's trash at roadsides, parks, and beaches.

Circle the pollution people can clean up in the park.

Children should circle the litter in the picture.

Concept: People can pollute ecosystems.
Pollution problems can be prevented or solved.

Workbook
THE GREAT OUTDOORS G19

Packing It Up

Many things people use are made of plastic or glass. And many things people buy come in plastic packages or glass containers. Both plastic and glass are useful materials. But there is a problem. These materials do not decay as paper, rags, and wood do. When plastic and glass things are put into dumps and landfills, they are still there many years later. So, to cut down on trash, many people **recycle** things made of plastic and glass.

Circle each piece of plastic and glass trash that people can recycle.

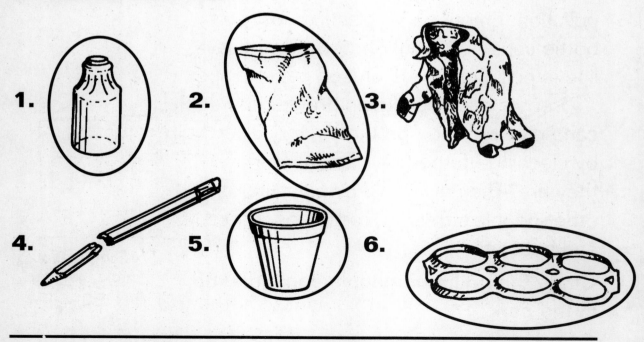

Name _____

Section 5
Lesson 3
Workbook

Second Life

Pollution is everywhere. It is in lakes, rivers, streams, and oceans. It is in forests. It is in deserts. And more and more, it is in rain forests. Pollution harms the land, water, and air in habitats around the world.

Many people recycle some of their trash to help solve pollution problems. They sort the trash into bins so it can be used again.

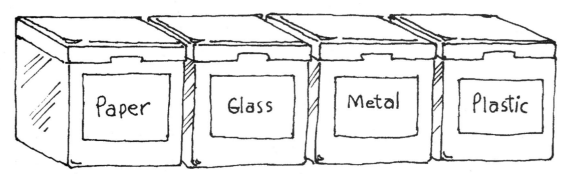

1. Draw an empty can in the correct bin.
2. Draw an old newspaper in the correct bin.

can: Metal bin
newspaper: Paper bin

Concept: People can pollute ecosystems. Pollution problems can be prevented or solved.

Workbook
THE GREAT OUTDOORS

Name _____

**Section 5
Lesson 4
Workbook**

Over and Over—Making Paper

This is a paper mill. The workers in this mill grind up trees to make wood pulp. Then they press the wood pulp into sheets of paper.

Old newspapers and brown bags are recycled. They are used to make more pulp. So, paper trash is used over and over again to make paper. You can make paper, too.

Read the steps below. Write the letter of the missing word in each blank.

A. iron   **B.** sheet

C. screen   **D.** water   **E.** newspaper

How You Can Make Paper

1. Tear old __E__ into small pieces.
2. Soak the pieces in __D__ until you have paste.
3. Spread the pulp on a __C__, and let it dry.
4. Remove the pulp, place a sheet on it, and __A__ it.
5. Peel the paper off the __B__.

G22   Workbook
THE GREAT OUTDOORS

Concept: People can pollute ecosystems. Pollution problems can be prevented or solved.

Name _____

Unit G Workbook

Language of Science

Fill in the missing word.

1. The natural home of a plant or an animal is _____ called a _____ **habitat** _____ .

Circle each word that names a habitat.

2. (desert) **3.** (pond)

4. flowerpot **5.** (rain forest)

6. Trees with needles and trees with broad leaves _____ can be found in a _____ **forest** _____ habitat.

7. A lake is larger than a pond but smaller than an _____ **ocean** _____ .

Language of Science (page 1 of 2)   Workbook THE GREAT OUTDOORS   G23

Name _____

Unit G Workbook

8. Lakes contain _____fresh_____ water.

9. Oceans contain _____salt_____ water.

Circle the word that tells about each picture.

10. (cactus) cacti   **11.** cactus (cacti)

Circle the words that complete these sentences about Jack.

12. Jack will _____ the cup.
      (recycle)   return

13. That will help solve _____ problems.
      party   (pollution)

SCIENCE AnyTime™
UNIT H: Show Time!

Section 1 • Art Gallery
- Lesson 1 • Color My World **H1**
- Lesson 2 • I See the Light **H2**
- Lesson 3 • Colors All Around **H3**
- Lesson 4 • Color Mix **H4**
- Lesson 5 • Spotlight on Color **H5**

Section 2 • Making Shadows
- Lesson 1 • What Makes a Shadow? **H6**
- Lesson 2 • The Right Moves **H7**
- Lesson 3 • Outdoor Shadows **H8**
- Lesson 4 • Shapely Shadows **H9**
- Lesson 5 • Light Blockers **H10**
- Lesson 6 • Bouncing Back **H11**
- Lesson 7 • Shadow Show **H12**

Section 3 • Making Music 1
- Lesson 1 • What Makes Sound? **H13**
- Lesson 2 • Shake It Up **H14**
- Lesson 3 • Making Noise **H15**
- Lesson 4 • Shaking the Air **H16**
- Lesson 5 • Bottle Music **H17**

Section 4 • Making Music 2
- Lesson 1 • Sounding Off **H18**
- Lesson 2 • Busy Beats **H19**
- Lesson 3 • Dueling Banjos **H20**
- Lesson 4 • Music Around the World **H21**
- Lesson 5 • All Together Now! **H22**

Language of Science **H23**

Name _____

**Section 1
Lesson 1
Workbook**

Color My World

One thing that makes the world an interesting place is the colors that we see around us. There are many colors, and they can look different at different times of the day.

You see colors because there is light. That is why you see colors better during the day, when there is more light, than you do at night, when there is less light.

1. Color the first picture above to show how things would look during the day.

   Colors should be bright.

2. Color the second picture to show how things would look at night.

   Colors should be dark.

Concept: Color is visible when light is present.
Colors can be mixed to form other colors.

Workbook
SHOW TIME!   **H1**

Name _____

**Section 1
Lesson 2
Workbook**

I See the Light

The brightness of the sun makes the outdoors seem more colorful. On a cloudy day, some of the sun's light is blocked by clouds. The things around us don't seem so colorful.

Indoors, the amount of light that falls on an object makes its colors seem brighter, too. We might need a lamp or a bright light bulb to see things clearly.

Draw something in each picture that would make the colors look brightest. Then color the pictures to show the colors you might see.

Possible responses:
1. lights, sun
2. sun
3. sun
4. lamp, sun through window

H2 Workbook
SHOW TIME!

Concept: Color is visible when light is present. Colors can be mixed to form other colors.

Name _____

**Section 1
Lesson 3
Workbook**

Colors All Around

If you wanted to paint a picture of what you see outside, you would probably need many colors. You would have to observe carefully the colors you see. Then you could mix some colors of paint together to make those colors. Many artists do this when they paint. After the artists decide what they want to paint, they decide which colors they will need.

1. What colors would you need if you were painting a forest scene?

Possible response: green and brown

2. What colors would you need if you were painting a snow scene?

Possible response: white and blue

Concept: Color is visible when light is present.
Colors can be mixed to form other colors.

Workbook
SHOW TIME!

Name _____

Section 1
Lesson 4
Workbook

Color Mix

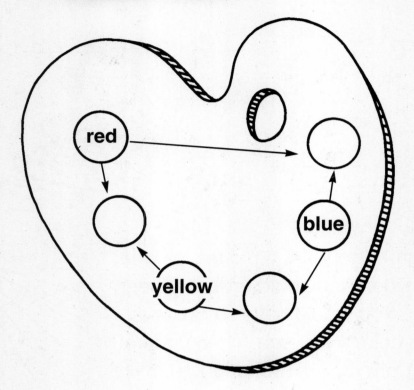

Artists need to learn which colors of paint to mix together to make just the colors they need. When they need green, they mix blue and yellow. When they need orange, they mix red and yellow.

If an artist needed purple, what colors do you think he or she would mix?

----
red and blue
_____

Color the palette above to show how to mix colors. Children should color the palette as the labels show.

H4 Workbook
SHOW TIME!

Concept: Color is visible when light is present. Colors can be mixed to form other colors.

Name _____

**Section 1
Lesson 5
Workbook**

Spotlight on Color

    Sometimes people want to change the color of something, but they can't paint the object. They <u>can</u> change the way its color looks, though. They do this by shining a colored light on the object. People who make movies sometimes do this with colored spotlights.

Color the pictures to show four different spotlights and the colors they have made the objects look.

The animal and spotlight should be the same color in each picture.

Concept: Color is visible when light is present.
Colors can be mixed to form other colors.

Name _____

**Section 2
Lesson 1
Workbook**

What Makes a Shadow?

If you look carefully, you can see shadows almost everywhere. Any object that blocks light makes a shadow. An outdoor shadow is made when an object blocks the light from the sun. Indoors, light may be coming from a lamp. If there is no light, there won't be a shadow. The object that made the shadow is always between the source of light and the shadow.

Draw shadows for the objects in the picture.

The shadows should be drawn in the direction away from the sun.

H6 Workbook
SHOW TIME!

Concept: Objects that block light cast shadows. Shadows change in position as the Earth rotates.

Name _____

Section 2
Lesson 2
Workbook

The Right Moves

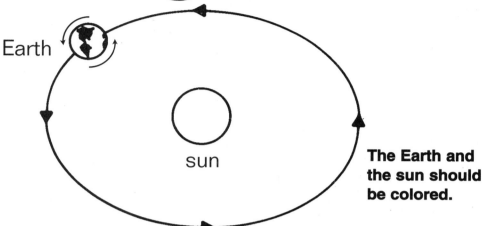

The Earth and the sun should be colored.

You have probably noticed that the shadow of a tree moves during the day. It seems to move from one side of the tree to the other. The shadow moves because the Earth rotates once each day. To us, it looks as if the sun is moving. Sometimes we even talk about the sun going up or coming down. But it is really the Earth that moves. As it moves, the shadows on the Earth change.

Write a sentence telling why outdoor shadows move during the day. Then color the Earth and the sun.

Possible response: Shadows move because the Earth rotates once each day.

Concept: Objects that block light cast shadows. Shadows change in position as the Earth rotates.

Workbook
SHOW TIME! H7

Name _____

Section 2
Lesson 3
Workbook

Outdoor Shadows

If you went outdoors and measured the same shadow at two different times, you would find that the shadow changes size. Why does this happen?

The size and position of a shadow change because the position of the sun changes. Look at where the sun is in these pictures. Look at the size and position of the shadows.

You can see that the size and position of the shadow changed as the Earth moved. **In these three boxes, draw your own changing shadows.**

1. **The three drawings should show a shadow that changes in size and position during the day.**	2.	3.

Workbook SHOW TIME!

Concept: Objects that block light cast shadows. Shadows change in position as the Earth rotates.

Name _____

Section 2
Lesson 4
Workbook

Shapely Shadows

Look at shadows around you. Is the shadow of a pencil the same shape as the pencil? Is the shadow of a bicycle the same shape as the bicycle? You will see that shadows have a shape that is like the object that makes them. Why do you think this is so?

When an object blocks light, the rays of light are stopped by the object. Other light rays go past the object's edges. The shadow is the area that the light rays can't get to.

In the picture above, color the toys that have the correct shadows.

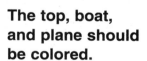

The top, boat, and plane should be colored.

Concept: Objects that block light cast shadows. Shadows change in position as the Earth rotates.

Workbook
SHOW TIME!   H9

Name _____

Light Blockers

Section 2 Lesson 5 Workbook

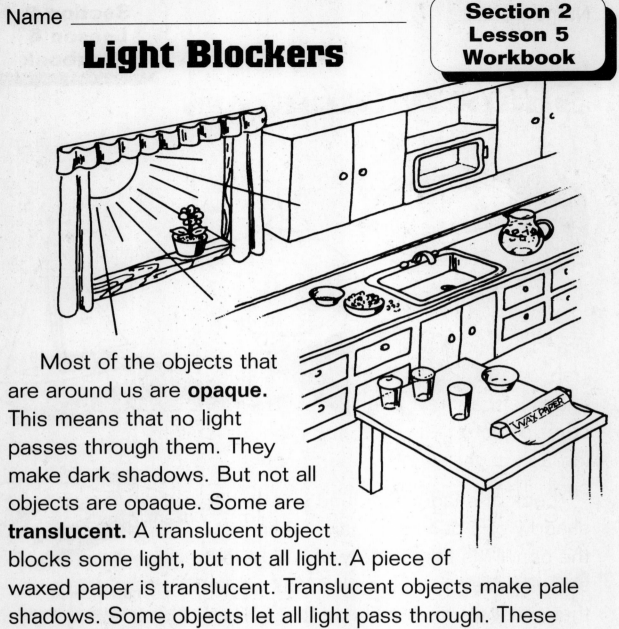

Most of the objects that are around us are **opaque**. This means that no light passes through them. They make dark shadows. But not all objects are opaque. Some are **translucent**. A translucent object blocks some light, but not all light. A piece of waxed paper is translucent. Translucent objects make pale shadows. Some objects let all light pass through. These are **transparent** objects. The clear glass in a window is transparent. Transparent objects do not make shadows.

1. Put an X on all the objects in the picture that are opaque.
2. Draw a box around the objects that are translucent.
3. Draw a triangle around the objects that are transparent.

box: waxed paper
triangle: glasses, pitcher, window glass
X: other objects

H10 Workbook **SHOW TIME!**

Concept: Objects that block light cast shadows. Shadows change in position as the Earth rotates.

Name _____

**Section 2
Lesson 6
Workbook**

Bouncing Back

We know that opaque objects make shadows because they block light. Transparent objects don't make shadows, because they don't block light. There are other objects that don't block light <u>or</u> let light through. Light hits them and bounces right back. These objects **reflect** the light.

One object that reflects is a mirror. When you look in a mirror to see yourself, you see your reflection. If you didn't have a mirror, what other objects could you use to see your reflection?

Make a list on the lines below.

Possible responses:

1. a shiny metal knife, fork, or spoon; a shiny plate or glass; foil; a
3. pool of water

2. _____

4. _____

Concept: Objects that block light cast shadows. Shadows change in position as the Earth rotates.

Workbook
SHOW TIME! **H11**

Name _____

**Section 2
Lesson 7
Workbook**

Shadow Show

Have you ever done a shadow puppet show? The puppets are behind a translucent screen that lets some light shine through. Only the shadows of the puppets can be seen. Usually the puppets act out a story.

Shadow theater was first performed many years ago in China and India. In China, puppets are often decorated and dressed in beautiful clothing, even though the audience can see only the puppets' shadows.

On this stage, draw your own puppets. Then write a title for your show.

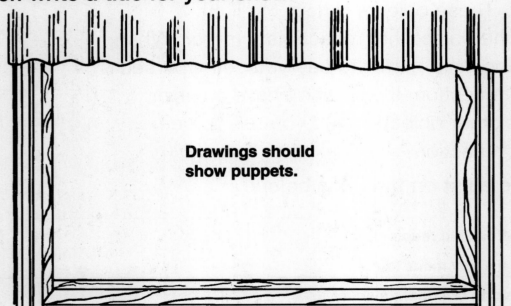

Drawings should show puppets.

Children should write a title for the puppet show.

H12 Workbook
SHOW TIME!

Concept: Objects that block light cast shadows. Shadows change in position as the Earth rotates.

Name _____

**Section 3
Lesson 1
Workbook**

What Makes Sound?

Every sound is caused by **vibration.** Vibration is a quick back-and-forth movement. When an object vibrates, the air around it also moves. This vibrating air enters your ears and makes your eardrums vibrate. That vibration tells your brain that you are hearing a sound.

In the picture above, circle all the things that would make sounds.

Possible circled objects: bell, people, jets, carnival rides

Concept: Sound is produced by vibrations.

Workbook
SHOW TIME! **H13**

Name _____

Section 3
Lesson 2
Workbook

Shake It Up

We hear many different kinds of sounds. One reason for this is that many different objects make the sounds we hear. The sound that is made when you hit a pillow is different from the sound of a bat hitting a baseball. The sound of a paper bag breaking is different from the sound of a glass breaking.

Draw lines to match these sound words with the pictures of the objects that make the sounds.

1. bang, bang

2. drip, drop

3. toot

4. slam

5. ring

6. snip, snip

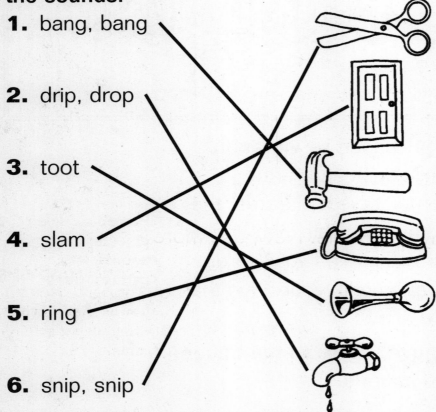

H14 Workbook **SHOW TIME!**   Concept: Sound is produced by vibrations.

Name _____

Section 3
Lesson 3
Workbook

Making Noise

You know that some sounds are loud and some are soft. That is because different amounts of energy are used to make the vibrations. The more energy that is used, the louder the sound is. Using just a little energy, tap a pencil softly on a table. Now, use more energy and tap the pencil harder on the table. Do you hear a difference?

On another sheet of paper, write about the band. Tell about the energy needed to play these instruments.

Possible response: The band uses a lot of energy when they play loud music.

Concept: Sound is produced by vibrations.

Workbook
SHOW TIME! H15

Name _____

**Section 3
Lesson 4
Workbook**

Shaking the Air

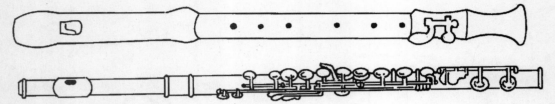

If you blow into a soda bottle, you can make a sound like a foghorn. You can hear the sound because blowing causes the air inside the bottle to vibrate. The vibrations make the sound.

Moving air makes the sound in some musical instruments, too. When people play a recorder or flute, they blow air into a hollow tube. The air vibrates, causing the sound.

Draw an instrument whose sound is made by moving air. Draw arrows to the places where the air vibrates.

> **Children should draw an instrument whose sound is made by moving air. Arrows should indicate areas of vibration.**

H16 Workbook **SHOW TIME!**

Concept: Sound is produced by vibrations.

Name _____

**Section 3
Lesson 5
Workbook**

Bottle Music

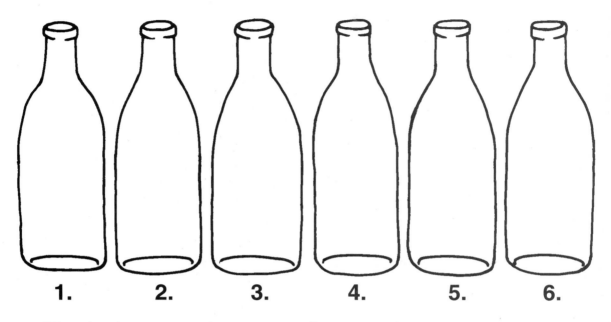

1.   2.   3.   4.   5.   6.

The highness or lowness of a sound is its **pitch.** The pitch of a sound is due to the number of vibrations each minute. The more vibrations there are each minute, the higher the pitch is.

When you pour water into bottles and then tap the bottles, you hear sounds. The less water there is in the bottle, the higher the pitch is.

Draw water in the bottles on this page. Draw the water so that if you tapped the bottles one after another, the pitch of sound you would hear would go from high to low.

Drawings should show the water levels going from low to high.

Concept: Sound can be controlled and described.

Workbook
SHOW TIME!   **H17**

Name _____

**Section 4
Lesson 1
Workbook**

Sounding Off

People make music when they control vibrations to make certain sounds. One way to make sounds is to make a sheet of some kind of material vibrate. People do this when they play a drum, a noisemaker, or a kazoo.

In the pictures below, color the part of the instrument that vibrates.

Children should color the membrane area of all three instruments.

1.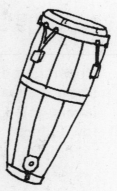
2.
3.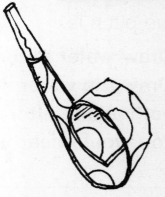

Workbook
SHOW TIME!

Concept: Sound can be controlled and described.

Name _____

**Section 4
Lesson 2
Workbook**

Busy Beats

When a drummer in a band or an orchestra plays a drum, he or she beats the drum in a pattern. That pattern of beats is called a **rhythm.** The rhythm, the pitch of the drum, and the energy the drummer uses to play the drum all control the vibrations that make the music we hear.

Write a sentence about how a drummer can make loud, fast vibrations.

Possible response: The drummer uses rhythm and energy to make loud, fast vibrations.

Concept: Sound can be controlled and described.

Name _____

**Section 4
Lesson 3
Workbook**

Dueling Banjos

Musical instruments like violins, guitars, and banjos have strings that can be made to vibrate. The player of the instrument controls the vibration of the strings. He or she can change the length or thickness of the strings to control the vibrations. The player could also change the sounds by tightening or loosening the strings.

Draw a musical instrument that has vibrating strings. Tell how the instrument is played.

> Drawing should show
> a stringed instrument.

Possible response: This instrument is played by pulling on the strings.

H20 Workbook
SHOW TIME!

Concept: Sound can be controlled and described.

Name _____

Section 4
Lesson 4
Workbook

Music Around the World

People all over the world make music. They use many kinds of instruments to make music. The instruments vibrate in many different ways.

1. colored: entire raspa Cuba	**2.** colored: bowl of bell England	**3.** colored: strings India
4. colored: entire gong Indonesia	Color the part of these instruments that you think vibrates.	**5.** colored: entire flute Japan
6. colored: strings Mexico	**7.** colored: skin of drum Nigeria	**8.** colored: strings United States

Concept: Sound can be controlled and described.

Workbook
SHOW TIME!
H21

Name _____

Section 4
Lesson 5
Workbook

All Together Now!

Instruments in an orchestra are grouped by the way they make their sounds. Drums are called **percussion instruments.** Percussion is the sound of one thing hitting against another. Violins are **stringed instruments.** They are played by pushing or pulling the strings and by making the strings shorter or longer. People play **wind instruments** by blowing into them. Blowing makes the air inside the instrument vibrate. The instrument vibrates, too. One wind instrument is a trumpet.

Draw some of the instruments in an orchestra. Label the instruments by using this key:
P = percussion, S = stringed, W = wind.

The instruments drawn by children
should be labeled according to the key.

Concept: Sound can be controlled and described.

Name _____

Unit H Workbook

Language of Science

Match the pictures with the words.

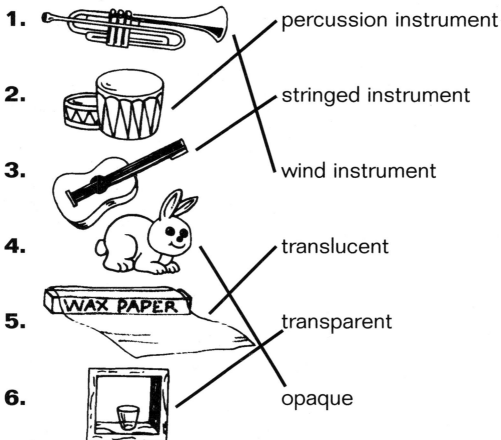

1. percussion instrument
2. stringed instrument
3. wind instrument
4. translucent
5. transparent
6. opaque

Circle the correct words.

7. Sound is caused by (shaking, (vibrations)).
8. (Vibration, (Rhythm)) is a pattern of beats.
9. Objects that ((reflect,) send) light bounce it right back.
10. (Rhythm, (Pitch)) is the highness or lowness of a sound.

Language of Science

Workbook SHOW TIME! **H23**